HAWAII

The Land and the People

by the world's greatest photographers

The port of Lahaina on Maui shelters a fleet of pleasure boats. Sailing boats, deep-sea fishing boats and glass-bottom boats can all be chartered here.

Dawn breaks over the ocean near Hana, on Maui's idyllic eastern coast.

Following Page:

Waikiki spreads toward the horizon, as seen from Diamond Head, Oahu's world-renowned landmark.

These members of the Konakeiki Dancers are all smiles at Kailua-Kona on the island of Hawaii. Kailua-Kona was once the capital of the Hawaiian kingdom.

HAWAII

The Land and the People

by the world's greatest photographers

GALLERY BOOKS
An Imprint of W. H. Smith Publishers Inc.

112 Madison Avenue New York NY 10016

THE IMAGE BANK®

111 Fifth Avenue New York NY 10003

First published in 1987 in New York by Gallery
Books, an imprint of W.H. Smith Publishers Inc.,
112 Madison Avenue, New York, N.Y. 10016

ISBN 0-8317-4386-7

For rights information about the photographs in
this book please contact:

The Image Bank
111 Fifth Avenue, New York, N.Y. 10003

Manufactured in Singapore

Produced by Robert M. Tod
Art Direction and Design by Mark Weinberg
Photo Editing and Graphic Coordination by George DuBose
Written by Rita Ariyoshi
Edited by Sheila Buff
Photo Research: Michelle McHugh, Laura Seid
Editorial Assistance: Elizabeth Loonan

"Paradise" is what most of the world calls Hawaii. The Polynesians, who first found this string of remote islands and named it for their legendary homeland Havaiki, called it in their chants "Ke Ao Nani," the beautiful world. The British explorer, Captain James Cook, who discovered Hawaii for the Western world in 1778, was more practical that poetic. He called Hawaii "The Sandwich Islands," after the Earl of Sandwich, who had largely financed his expedition. Mark Twain called Hawaii "The loveliest fleet of islands that lies anchored in any ocean." Today, more than five million tourists a year call Hawaii their "dream trip," while one million people call it home and refer to the islands affectionately as Hawaii Nei, our Hawaii.

The 132 islands that form the Hawaiian archipelago are, in reality, the tops of mammoth sea mountains that rise from the ocean floor in the middle of the vast Pacific. The land drops off precipitously right offshore, plunging to abyssal depths that harbor an astounding assortment of marine life.

The sea mounts were formed by submarine volcanic eruptions along a fault in the ocean's floor known as the Pacific Ring of Fire. The activity took place in the quiet of the deep for millions of years, the lava and basalt outpourings building huge, mounded pillows, until the first land emerged above the waves about twenty-five million years ago. The volcanoes are still building land, with Mauna Loa and Kilauea on the island of Hawaii erupting frequently in spectacular outbursts.

The new volcanic islands were sterile and lifeless. Nothing moved but the wind, and nothing grew except the land itself. The seeds of life arrived in the ocean currents, in the jet streams and with migrating birds, at the rate of one new species every fifty thousand years. In isolation, these colonizing plants and birds evolved into one of the most fascinating ecosystems in the world. Most of Hawaii's native flora and fauna are found nowhere else on the planet.

Due to the introduction of more competitive exotic species and the ignorance and manipulation of man, many of these plants and animals have become extinct or endangered. From today's more enlightened view of nature, efforts are being made to save what is left of the fragile environment. Wilderness preserves have been established on most islands. Much of this pristine beauty is accessible to the public.

The first people to arrive in Hawaii were Polynesians from the Marquesas Islands, some 2,500 miles to the southeast. They sailed in great voyaging canoes, steering by the wind, the stars and the ocean currents, arriving in Hawaii more than a thousand years ago. The next great wave of settlers sailed from Tahiti about A.D. 750. These first Hawaiians lived in harmony with their limited island environment and developed a philosophy and culture that has influenced every facet of modern Hawaiian life, even though the Hawaiian people themselves now number less than ten percent of the total population.

With the arrival of Captain Cook in 1778, everything was forever changed for Hawaiians. It was as if a spaceship had landed. The introduction of diseases to which they had developed no natural immunities decimated the population. The new concept of buying and selling their land left them without any. Vast plantations of pineapple and sugarcane carpeted the plains and hillsides. Plantation owners, due to the decline of the Hawaiian population, imported laborers from China, Japan, Portugal and the Philippines. The Hawaiian culture was soon eclipsed, and their queen deposed and imprisoned in her own palace.

The miracle of it all is that so much has survived. It is the thread of the Hawaiian culture that binds together the many aspects of modern Island life. The beautiful philosophy of aloha, of welcoming the stranger with love, of greeting each person with the knowledge that each stands in the presence of God as they speak, has paradoxically spelled the end of the ancient way of life, and made the new way wondrous — and very Hawaiian. Hawaii is a model of race relations in a world torn by racial and religious strife. Interracial marriages number over fifty percent and no one is in the majority. The many cultures, while retaining their distinct flavors, have blended into one happy "chop suey" family, with everyone celebrating everyone else's holidays and customs. Islanders today are becoming a new race, a golden blend of East and West.

With the election in 1986 of John Waihee as governor, a native Hawaiian took the helm of government for the first time since the reign of Queen Liliuokalani. In his inauguration address, given on the grounds of Iolani Palace, Governor Waihee was clear in his pronouncement that his election was not at all a victory of Hawaiians, but a victory of aloha, of ohana, of all the family. The spirit of the occasion was perhaps best personified by a pretty Japanese-American girl dancing the hula in her kimono while the United States Marine Band waited in the wings for their cue to strike up a Sousa march.

HAWAII

It is on the island of Hawaii, the newest island in geological terms, that old Hawaii lives. In the fire and steam of the volcanoes, in the stark drama of the lava-walled temples, in the hearts and minds of the people, the ancient ways survive.

Called the Big Island, with 4,038 square miles, Hawaii makes up almost two-thirds of the land area of the state. Its topography is the most diverse in the state, ranging from the tropical beaches of Kona to the snow-crowned heights of the mountains, so that it is possible to swim and ski, all in a day. The highest mountain on earth is here, Mauna Kea, rising more than thiry thousand feet from the ocean floor, though its summit stands at only 1,796 feet above sea level. In sheer bulk, Mauna Loa is the largest mountain in the world. Kilauea ranks as the world's most active volcano, awake more often than it is asleep. The only commercially grown coffee in the United States is raised on this remarkable island, and the country's largest privately owned cattle ranch, Parker Ranch, is tucked away here in this westernmost part of the American West.

Hawaii's greatest king, the man who would unite all the Hawaiian islands into one nation, Kamehameha I, was born on the island of Hawaii in 1758, on a night when Halley's comet streaked across the Hawaiian skies. Alapai, king of the island, ordered the infant killed at birth because a priest prophesied that he would be a "killer of chiefs and ruler of all the islands." Kamehameha's mother fled with him and hid in the mountains of Kohala, raising him in secret. Alapai's fears were well founded, for when the young Kamehameha reached manhood, he gathered an army and defeated him.

The birthing stone where Kamehameha was brought into the world is in Kohala. The original statue of the king, the model for the famous copy that stands across from Iolani Palace in Honolulu, presides over the little Kohala town of Kapaau. Ahuena Heiau, the temple where the great conqueror worshipped his terrible war god, Kukailimoku, has been completely restored. It sits on a spit of land that juts into Kona Bay and belongs to the hotel King Kamehameha.

It was in Kealakekua Bay, not far away, that Captain James Cook was killed in 1779. The British captain had sailed into Kealakekua with his two ships, *Resolution* and *Discovery*. The strangers arrived during the Makahiki, the harvest festival, and were warmly received by the people. They were treated like royalty and offered the bounty of the land. It is said that they overstayed their welcome, however, and the Hawaiians were glad to see them leave. The English ran into a storm at sea that dismasted one of their ships, forcing them to return to the shelter of the harbor for repairs. By this time, the Makahiki with its imposed period of peace was over and their weary hosts were not as cordial. Provisioning the ships had put a strain on the local subsistence economy. Trading disputes broke out between Cook's crew and the local populace, culminating in the theft of a British rowboat, probably for the iron in its nails and fittings. Cook and his men decided to go ashore, kidnap the chief and hold him until the Hawaiians returned the boat. In the battle that ensued, Cook was killed. Some say this decisive blow came from a war club wielded by Kamehameha. A monument to the intrepid explorer, who first brought news of Hawaii to the rest of the world, now stands at Kealakekua Bay.

The other figure who looms tall in any discussion of the island of Hawaii is Pele, the goddess of the volcano. Legend states that she came from Kahiki, now translated as Tahiti, although the original word meant any foreign place. She arrived in a canoe with her divine brothers and sisters, trying each island of the chain and finally making her home on the Big Island at Kilauea Volcano in the pit called Halemaumau. Of all the deities of the Hawaiian pantheon — the gods of the ocean, sea and wind, the gods of the harvest and fertility, Ku, Kane, Kanaloa, Lono — Pele alone survives. Many people, even the type not usually given to flights of fancy, say they have seen her. Sometimes she is young and beautiful—with piercing, coal-black eyes and ehu hair, the color of flame. Other times she is an old woman who walks on lonely roads, with only her white dog for a companion. Usually she is seen right before a volcanic eruption, before the seismographs at the volcanic observatories begin to register the stirring of magma deep within the earth.

It is easy to imagine her presence at Hawaii Volcanoes National Park. She is in the sulphur fumes, in the steam rising in ghostly forms from the vents of the volcano. She stalks the hiking trails of the caldera, where the ground is still warm. No one eats the delicious, bright-red ohelo berries that grow in the park without offering the first of the pick to Pele, and no one takes the lovely ohia lehua, her sacred flower, to tuck casually in their hair. Those who, in ignorance, have removed souvenir pieces of lava from her land claim to have nothing but ill fortune until the rocks are returned.

Pele's fiery outbursts can be seen from the town of Hilo, the county seat of the island.

With the domain of the temperamental goddess at their backs, and facing a sea that has often turned hostile, Hilo's citizens have had more than their share of tragedy. On April Fool's Day, 1946, a tsunami, or series of seismic shock waves generated by an earthquake, slammed into Hilo, sweeping across the town. When it receded, it took half the town with it.

The city was rebuilt, this time with a stone breakwater standing guard across the mouth of the bay. The new wall was no match for the tsunami that swept in from the ocean in 1960, again destroying the entire waterfront area of the city.

This time Hilo surrendered and retreated for good, turning the waterfront into a park — a very beautiful park. Liliuokalani Gardens with its torii gate and stone lanterns reflects the Japanese heritage of Hilo.

Japanese immigrants poured into Hilo in the late nineteenth and early twentieth century, seeking work on the sugar plantations that stretched inland from the Hilo and Hamaku coasts. They stayed, raised their families and became part of the emerging Hawaiian society. The *Hilo Times* newspaper is still published in Japanese. The daily (except Sunday) Suisan fish auctions seem to sum up what people love best about Hilo. The action is loud and the bidding conducted in a glorious, esoteric blend of Japanese, pidgin, Hawaiian and English. The auctions have become something of an "insider" tourist attraction.

Other popular Hilo stops are the Lyman Mission House and Museum, where the New England missionary Lyman family has restored its original home and collected a fascinating melange of natural history exhibits and historical artifacts, and the Mauna Loa Macadamia Nut Mill and Orchard with its factory tour and free nuts.

Gardens abound around this often misty town. Orchid and anthurium nurseries export their blooms to other islands and colder climates. The orchid industry is, in fact, the largest in the world.

Two of Hawaii's prettiest waterfalls are close by: Rainbow Falls, named for the aura that dances across its cascade in the mornings; and Akaka Falls, a 442–foot plunge into a fern and ginger-rimmed pool.

Beyond Hilo, the 2,000-foot cliffs of the Hamakua Coast loom majestically. Waipio Valley slices into the coastal fortress, deep, green and lush, with the twin falls of Hiilawe dancing in the distant reaches. Waipio was a political and cultural center of old Hawaii, with thousands of people raising taro in the rich soil. The river was so full of fish, it is said they could be caught with bare hands. Today, a few tenacious farmers remain in this beautiful, isolated place, still raising their taro. As for the fish, like the people, they have diminished in number.

By following a rugged trail that starts near the black sand beach, the determined explorer may reach neighboring Waimanu Valley, where no one lives and the peace is completely undisturbed.

Much of the Big Island is characterized by solitude, and by a sense of time that seems to have stopped and then become encapsulated, while the tumult of progress whirls on the periphery.

Small plantations towns go about their business as usual, and the great cattle ranches of the interior graze their herds on the undulating emerald hills. There are round-ups and rodeos, macademia nut festivals and Kona coffee queens.

Richard Smart, heir to the vast Parker Ranch, was the first of the Big Island cattlemen to let go some of his holdings for resort development. Now the whole Kohala area is sprouting luxury resorts and manicured golf courses. Boutiques sidle up to the country stores and leather shops in the ranch town of Waimea.

Traditionally, the playground of the Big Island has been the sunny Kona Coast, centered around the town of Kailua-Kona. Here, the sea offers some of the best big-game fishing opportunities in the world, with record marlin the norm. Snorkeling, diving and sailing enthusiasts take advantage of the clear, clean waters without regard to seasons.

The Kona Coast is not all sun and fun, however. It is here that Hawaiian history is most visible. Hulihee Palace, overlooking the waters of Kailua Bay, was once the holiday retreat of Hawaiian royalty. Directly across the street, Mokuaikaua Church is the oldest Christian church in the Islands. Outside of town, Saint Benedict's, the famed Painted Church, rests on the slopes of Mauna Loa. Below, right on the edge of the ocean, is the best preserved of the ancient Hawaiian temples, the Puuhonua o Honaunau, now a national park. Fearsome tiki images stand as sentinels. The temple, whose thick walls are still standing, was a place of refuge in the old days, offering sanctuary to women and children in time of war, and offering a way of atonement for those accused of crimes.

All across the Big Island, the island of flowers and flames, are the contrasts of time, with ancient petroglyph fields strewn across hotel grounds, and the old pathways of kings traversing golf courses. Over it all is the specter of Pele, the remnant, the holdover, the one who refuses to go away, who regularly and with awesome force reminds people of her power.

OAHU

Kamehameha, the first chief to unite all the Hawaiian islands into one kingdom, landed with his armada of war canoes at what is now the Waialae Country Club, scene of the annual Hawaiian Open golf tournament. He moved his troops inland and fought the decisive battle for control of Oahu at the Nuuanu Pali, just above the place where the modern Pali Highway tunnels through the Koolau Mountains. The defeated warriors were forced over the precipitous cliffs.

Even at that raw stage of its development, the collection of grass houses clustered around the muddy waterfront of Honolulu was considered much too urban by the victorious king. With his entourage, Kamehameha sailed back to his native island of Hawaii, where he ruled until his death.

People still complain that Oahu is much too metropolitan. Traditionally called "The Gathering Place," the population of the 607.7–square-mile island has topped 800,000, representing more than eighty percent of the state's population. The city of Honolulu sprawls in all directions, even out to sea, for its legal boundaries encompass the entire northwestern chain, stretching 1,091 miles from Nihoa to Kure Atoll. The islands of Midway are the exception. This group is not a part of the state of Hawaii and is administered by the federal government. Most of these northwestern islands are a national wildlife refuge.

The skyscrapers of Waikiki and downtown taper off to suburban homes that cluster in the valleys and cling to the sides of the mountains like limpets to rocks. The island is laced with interstate highways that cannot be linked to any other state.

With its two natural deepwater harbors, Honolulu Harbor and Pearl Harbor, Oahu has been a magnet for opportunists, entrepreneurs and armies. The government has been centered in Honolulu since King Kamehameha III moved his capital from Maui to Oahu in 1843.

It was the obvious strategic importance of Pearl Harbor that earned Hawaii its first sugar treaty with the United States, giving rise to the plantation era and the way of life still pursued in much of rural Hawaii. It was also Pearl Harbor that drew the fire of enemy bombers on the morning of December 7, 1941, triggering America's entry into the war in the Pacific. Pearl Harbor, with its major draw, the *Arizona* Memorial is, today, Hawaii's number-one visitor attraction.

An anomaly in the middle of the ocean, Honolulu builds its towers skyward as if they could rival the grandeur of the brooding green Koolaus that rise in misted ramparts beyond the city. Depending on perspective, they sometimes succeed.

Honolulu is the commercial and cultural crossroads of the Pacific. The airport is one of the busiest in the nation, just behind New York, Washington and Chicago. The state has consciously sought leadership in Pacific affairs. The East-West Center at the University of Hawaii has brought together scholars, teachers and leaders from East and West. All of its programs are designed to foster international understanding, especially between Asian-Pacific peoples and Americans. Hawaii itself serves as a living laboratory for the study of cross-cultural sensitivities.

Even with the huge military spending that is lavished on Oahu, the amount of land devoted to agriculture, and the bustling pace of downtown Honolulu, the business of the island is play. Tourism is the primary economic force, touching almost every facet of Island life.

Waikiki, the jewel in the crown of Hawaiian resorts, has grown from a languid tropical strand of sand with a couple of hotels and a lot of palm trees to a sophisticated, cosmopolitan resort city on the order of Rio de Janeiro, in less than forty years.

The giants of the hotel industry are all here: Sheraton, Hilton, Regent, Hyatt, Westin and a host of smaller chains. They offer visitors accommodations at every budget level, with the price of the room rising in direct relation to its proximity to the ocean. Some hotels have marble hallways, glass elevators and attendants, while others resolutely preserve traditional lauhala wall coverings and chandeliers made of seashells.

Restaurants reflect the discerning tastes of both a local populace that demands excellence in food preparation, and a sophisticated legion of travelers accustomed to the best the world has to offer.

Treasures and baubles of every description and origin are gathered in Wakiki, everything from Japanese antiques to Irish crystal. The Paris designers are here and so are the Australian sheepskinners.

Waikiki has a fine little zoo with evening concerts in the summer. It's billed as the wildest show in town. There's also a notable aquarium. Kapiolani Park, with its venue of festivals, fairs and shows — including the Kodak Hula Show that's been running for fifty years straight — even has a golf course (the Ala Wai) right on the edges.

The big attraction, however, will always be the beach, that fabled strip of sand glistening between the rows of hotels and the sun-dappled ocean.

The extinct volcano Diamond Head presides over the scene, its familiar profile giving everything the look of a picture-perfect postcard.

Beach boys, who have worked the sands since the days when the Matson luxury liners used to deliver royalty and celebrities to the "Pink Palace of the Pacific," the Royal Hawaiian Hotel, still give surfing lessons to grandmothers and take tourists to ride the waves in outrigger canoes. There are also catamaran rides, aquabikes, jet skis and almost every water toy a person could desire.

After dark, Waikiki glitters. The hula dancers move in time-honored motions, but they wear sequins and cellophane skirts. The luaus lure their guests, but they serve macaroni salad and fried chicken along with pig and poi. Vegas-style dinner shows, discos and lounge acts light the night. Steady streams of people meander along the sidewalks, sometimes moving faster than the traffic. Music spills from doorways, and streetside stalls hawk everything from gold chains sold by the inch to coconut-shell ashtrays.

Somehow, despite the whirling pace, the cacophony of sound and the crowds of people, Waikiki still has a magic. It still enchants. Couples who haven't held hands in years stroll arm in arm, she with a flower in her gray hair and he with his gabardine pants rolled up. Honeymooners come by the millions from Tokyo, Toronto, New York, and all points in every other direction. What's more, they keep coming back, anniversary after anniversary, often requesting the same room. One hotel is unable to throw away its faded chairs. They have to keep them in storage and drag them out for returning couples who "want everything in room 807 to be the same." Another hotel warehouses tropical wardrobes for its regular guests so they don't have to pack. When they arrive, their clothes are hanging neatly in the closet.

Maybe it's the moonlight that imparts the magic. There's nothing quite like it when the moon sails full and silver over Diamond Head and etches soft paths of light in the ocean. Maybe it's the heady scent of flowers carried on the tradewinds, acting as more of an aphrodisiac than any French perfume. Whatever it is, Waikiki has it.

Aside from the premier resort itself, Oahu abounds in attractions, both natural and man-made. The number-one paid visitor attraction is the Polynesian Cultural Center. Here, on 27 lagoon-laced acres, it is possible to simulate an entire South Seas odyssey all in one day. Owned and operated by the Church of Jesus Christ of Latter Day Saints, the Center was developed to provide scholarships for Pacific island students to attend the Brigham Young University Hawaii campus. They share the unique dances, songs, arts and customs of their far-flung homelands in Broadway-quality shows and spontaneous expressions of caring and humor. There are seven "villages" bordering the network of lagoons. Canoes carry visitors to Tahiti, Tonga, Samoa, Aotearoa (New Zealand), Fiji, the Marquesas and, of course, Hawaii.

Another attraction, Sea Life Park, sits snugly at Makapuu Point where the mountains meet the sea. This marine park has a magnificent setting, a 300,000-gallon natural reef tank, dolphins that dance the hula, a whale show, seals, penguins and a scrimshaw museum.

At Waimea Falls Park, lovely Waimea Falls cascades into a natural swimming pool, inviting bathers and picnickers to linger. The historic valley has become a botanic enclave of rare tropical plants, all clearly labeled and set in lovely gardens. There are walkways and also guided tours by tram.

The history of Hawaii is visible in the buildings left behind by kings and preachers. Iolani Palace and the Queen Emma Summer Palace were the domain of the former, while the latter lived in the saltbox homes now known as Mission Houses Museum.

The Bishop Museum, with its full schedule of Hawaiian activities and its planetarium, displays a vast collection of Pacific artifacts.

Because of all the activities and attractions, people hardly notice just how beautiful Oahu really is. The view from the Pali Lookout, with windward Oahu spread out below and the towering Koolaus parading into the distance, has to be one of the most glorious in the world.

Twenty minutes in a straight line from downtown are two of the best beaches in the state, Kailua and Lanikai, crowded with local people on weekends and just about deserted the rest of the time. The whole windward coast, with its placid bays and country ways, is a visual treat.

The North Shore is where monstrous winter waves tempt surfers from around the world to test their skills and their spirits against the sea. When not riding the waves, surfers gather in Haleiwa, an old plantation town turned trendy with boutiques, art galleries and seafood restaurants.

At Makaha, the waves break closer to shore and the surfers' exploits are easier to admire.

Hanauma Bay, formed by the erosion of an ancient volcanic crater, is a marine life sanctuary teeming with jewled fish — a snorkler's heaven. A piece of bread will buy a lot of finned friends here.

Acres of pineapple and waving emerald seas of sugarcane dominate the central plains of Oahu, impressing a rural charm on an island that sometimes looks wistfully over its shoulder at days gone by, while never missing a step in the forced march forward.

MAUI

Maui is called the Valley Isle because of the cleft between the two great volcanoes that form the island, Haleakala with its desolate moonscape crater on one side, and older Mauna Kahalawai, commonly known as the West Maui Mountains, draped in rainbows and mist, on the other.

The topography of Maui symbolizes the split personality of the island, very definitely divided, yet forming a harmonious and pleasing whole. Hana and Kahakuloa are the old Maui, where life is lived close to the elements and offerings to the ancient fish gods are still left at shoreline shrines. The new Maui has been called the Marin County of Hawaii. From the lofty heights of Haleakala, in the area called Kula, transplants and transients from Mainland U.S.A. have set up A-frame housekeeping, complete with center-island kitchens and redwood hot tubs.

Maui's fabulous coastal resorts have become the playground of the world and are attracting the rich and famous. The island's diversity is a good part of its charm.

The island is named for the ancient superhero of Polynesian myth, Maui. Said to be born of a mortal mother, he was adopted by the gods of the ocean and raised in their watery realm. The child, however, was undisciplined, and used the powers the gods bestowed on him for nothing but mischief. In divine exasperation, the gods finally returned him to his mother. Maui is said to have lassoed the sun, and forced it to travel more slowly across the sky, awarding this part of the world longer and warmer days.

Blessed with the sunshine and cooled by gentle tradewinds, Maui has become a mecca for people seeking respite from colder climates and tumultuous times.

They follow in noble footsteps. Kaanapali, now a model resort sprawling along a crescent of gleaming sand, was once the playground of Hawaiian royalty. With an enormous retinue, they would travel from their capital at Lahaina to Kaanapali for holidays of feasting, dancing and surfing on the gentle, rolling waves. They'd race their outrigger canoes through the surf and fish from the abundant waters. Where golfers now putt along the greens of the Royal Kaanapali golf courses, the Hawaiians played ulu maika, a form of lawn bowling. In the absence of the present boutiques, they spent languid afternoons making soft tapa cloth from the bark of the wauke tree. They dyed and decorated the fabric, and fashioned it into clothing and bed linens.

Having catered for years to the demands of pampered sovereigns, the tourism industry is a breeze for Mauians.

The Wailea-Makena area is the latest to assume the royal tradition of hospitality. There are three championship golf courses, five wonderful beaches and a string of grand hotels.

The first resort on Maui, in fact the first Hawaiian resort outside Waikiki, is at Hana, which is, paradoxically, the most traditionally Hawaiian part of the island.

Queen Kaahumanu, the favorite wife of Kamehameha the Great, was born here. In the 1790s she was called by the British Captain George Vancouver "The most beautiful woman in the South Pacific." In ancient times, Hana seethed with royal intrigues and raged with warfare. The largest heiau, or temple, of the Hawaiian faith stands at Hana. Its tumbling lava walls are a grim reminder of the severity of the "old time religion," and of the fact that even gods have their season.

Only small planes and intrepid drivers reach Hana. It sits at the end of the winding, ambitiously named Hana Highway, a ribbon of road that hugs the cliffs and dips into fern-choked valleys. It has 54 one-lane bridges, countless magnificent vistas, several waterfalls and ripe guavas that hang seductively over the road.

Little detours yield star-wish rewards — like Keanae peninsula. Besieged by surf, its ragged shoreline juts defiantly into the ocean that once overwhelmed it when a tidal wave slammed ashore at five hundred miles an hour in 1946. Only the old stone church stood. Now quilted in taro patches punctuated by two old churches and a few small homes, life at Keanae proceeds as it has done for centuries — slowly. There is a reverence for the land and the sea, born of total dependence on both.

Keanae is a prelude to Hana — it prepares the way.

The little town at the end of the road was called by the ancient Hawaiians, "the land of mist and low-lying cloud." Later arrivals have simply, tersely, dubbed it "Heavenly Hana." Famed aviator Charles Lindbergh, perhaps wanting an edge on Paradise, chose to spend the last months of his life here, and is buried beside a little church in neighboring Kipahulu. Hana's other most-visited sites are the hodgepodge Hasegawa General Store, which sells everything from tractor parts to sunscreen number fifteen, and the Seven Pools of Kipahulu, a series of waterfalls and pools that channels rain from the heights of Haleakala to the sea.

Those lofty heights cradle the largest volcanic crater in the world, big enough to swallow Manhattan, skyscrapers and all. America's astronauts trained here, atop Haleakala, for their first moon landing, so akin is this eerie caldera to the face of the moon. It is mounded with giant cinder cones in shades of russet, amber and olive. Many of its birds and plants are found no place else on the planet. The hauntingly beautiful silversword, with its pale and shining leaves, appears to be a flower of another world, whose seeds might have come with a long-ago comet.

The middle slopes of the mountain are pure Marin. Even in the old days it was called Nu Kaliponi — New California, because the fertile fields grew the vegetables that fed Forty-Niners during the Gold Rush.

At its cooler elevations, Haleakala still grows a cornucopia of crops for Hawaii's tables. Even the noble grape does well. Maui brut champagne from the Tedeschi Winery was served at President Reagan's second inauguration.

This is also ranch country, with herds of cattle grazing the high mountain pastures. The annual Makawao Rodeo on the Fourth of July is the biggest in the state. Then Maui's cowtown, nicknamed "Macho-wow," leaps to life with parades, bunting and championship riding and roping.

Conversely, Haleakala comes festooned with flowers, dressed in acres of sweet-smelling carnation and the wondrous African import, the protea flower. Maui blossoms are shipped around the world.

On an island full of surprises, perhaps the biggest is Lahaina. It is a genuine New England village, complete with white clapboard buildings, widows' walks and weathervanes, set among the palm trees.

Once the royal capital of the Hawaiian kingdom, Lahaina also served as the center for the Yankee Pacific whaling fleet. As many as fifteen hundred sailors at a time wandered along Front Street, carousing in the grog shops and searching for willing women. Their sins aroused the righteous wrath of a group of New England missionaries, who had come to bring salvation to the Hawaiians. They were already well entrenched with the ruling families and succeeded in having laws passed prohibiting almost everything except the Psalms. The battles between the missionaries and the sailors became legendary.

Lahaina today looks very much as it did in its lusty heyday. Many acres of the town have been declared a National Historic Preservation District.

The main area of Lahaina is anchored on one end by the famed Pioneer Inn, which presides over a harbor of pleasure craft, and on the other by the largest statue of Buddha outside Asia, at the Jodo Mission.

Ironically, the whales still return to Maui, having barely survived the decades when they were hunted. They breed in the warm, sheltered waters during the winter months, giving birth to two-ton babies. Their fantastic gymnastics, staged right offshore, delight Mauians and amaze visitors.

The surprising and very separate parts of this island are actually facets of the same diamond. Their luster depends on the light that is shed upon them.

KAUAI

When Hollywood needs a Garden of Eden, it casts Kauai. The island has been the setting for films as diverse as *South Pacific* and *King Kong*. It is probably the closest place on earth to a vision of Paradise.

Kauai is geologically the oldest of the major Hawaiian Islands, and the fourth largest. It was formed by the volcanic activity of Mount Waialeale. The old mountain is the wettest spot on earth; its hoary, 5,000-foot summit is rarely seen. Cloaked in gray clouds and veiled in mist, Waialeale cradles the Alakai Swamp, where 100 inches of rain fall each year. The inhospitable Alakai has become the last refuge of many of Hawaii's rare birds. Some species, once thought to be extinct, have been spotted in the remote forests. Others are known to be so few in number, and their habitat so limited, that it is only a question of time until their song is gone forever.

While people, just by their presence, may have been unkind to Kauai, time has done its work well on the island, sculpting the mountains into deeply gorged and jagged gems. The ocean has gifted it with gleaming coral sand, strewing it in secluded coves and along great uninterrupted stretches of beach. The rains of Waialeale have fed the broad rivers and tumbling waterfalls for centuries. The island is understandably well-endowed with the richest green and adorned with beautiful flowers.

Her handsome kings loomed larger than life and were the focus of some of the most fascinating romantic intrigues in Hawaiian history and myth. Lohiau managed to capture the fancy of Pele herself, and paid with his life, for Pele is, above all, a jealous goddess. Kaumualii, who had refused to surrender his island to the might of the Kamehameha army, was, in the end, seduced by the great warrior's favorite wife, Kaahumanu.

Kauai was the first island to be visited by Captain Cook, who sailed into Waimea Bay, and the last island to join the united Hawaiian kingdom. The first sugar plantation in the islands was started on Kauai, and Kauai's tiny neighbor, Niihau, a part of Kauai County, is the last place left on earth where the Hawaiian language is the primary tongue. Privately owned Niihau is called "The Forbidden Island," since no one is permitted to go there without invitation, and it is understood that no one will be invited. Here, native Hawaiians are free to pursue their own way of life, speak their language, and practice their ancient culture without fear of exploitation or of becoming the object of tourists' curiosity. The Robinson family, owners of Niihau, have applied for permits to land helicopter tours on a remote part of Niihau. Tourists, according to the family, will not be permitted to leave the landing site and will have no contact with the people of the island, but they will be able to boast that they have been to Niihau.

Niihau islanders are famous for their exquisite shell leis, which sell for hundreds, even thousands of dollars, depending on the coloring of the tiny lustrous shells, the style in which they are arranged, and the number of strands. Aside from this tangible expression of Polynesian art, Niihau remains a mystery.

Like their Niihau neighbors, the people of Kauai have always marched to a different drummer. They cherish their rural way of life centered around the plantations. Most of the people live in small towns with populations of under four thousand.

Tourism is making great inroads on the island, bringing with it both the attendant problems and prosperity. The encounters between developers and residents have not always been cordial, for Kauaians have a history of fierce independence. In their struggle to open the island to new economic opportunities for their children, and to maintain what is precious from the past, they have exacted small victories with large implications, like the stipulation that no building may be constructed higher than a palm tree.

There are now four principal resort areas on the island: the county seat of Lihue and its harbor, Nawiliwili; the Wailua–Kapaa stretch of coast; Poipu Beach on the southern side; and lush Hanalei.

It is impossible to circle Kauai by car because of the fortress-like cliffs called Na Pali. Much of the spectacular beauty of the island is inaccessible except by helicopter tours, rafting or determined hiking.

The ramparts of the Na Pali Coast can be approached by boat only in mild weather. Besieged by surf, this rugged beauty has beaches and rivers, breathtaking sunsets and solitude to spare. There are waterfalls in the valleys and ripe fruit for the picking. There are legends of lost tribes and half-mad men fleeing civilization. The isolation breeds such stories, particularly around a campfire with only the stars for sentries and the deep, brooding valleys for companions.

It is possible to peek into the broad Kalalau Valley from the heights of the Kokee lookout. Here the towering cliffs reach four thousand feet and the green of the valley spills into the sea.

Waimea Canyon, which has been ambitiously, but not altogether inaccurately, called the Grand Canyon of the Pacific, can also be viewed from a vantage point just below Kokee. As far as the eye can see, the rocky crags of Waimea stretch to the horizon in shades of rose, amber and ochre. They stretch back into millions of years of time, as the waters from the Alakai River slowly, patiently carved the canyon, etching it, shaping it and exposing it. The wind tunnels through the deep gorges, sometimes whispering, sometimes roaring, but always present. It is the abode of wild mountain goats and boars. It belongs, by association, to Arizona and Colorado, not Hawaii. Its vastness, its time-worn cliffs and pinnacles, its roseate coloring are a complete surprise in the middle of the Pacific.

Of all the legends of Kauai, the most persistent are the tales of the Menehune. Akin in traits to the leprechauns of Ireland, these little people are said to have inhabited Hawaii before the Polynesians. They are described as perhaps only two or three feet tall, with extremely strong arms and legs and protruding stomachs. They feed on taro tops, sweet potatoes and haupia, a smooth coconut pudding. Their favorite food, in the old days, was shrimp, and it was also their medium of exchange. A Menehune laborer received a shrimp for his daily wage. The little people were nocturnal workers, completing their projects in a single night. They are credited for building a network of fishponds and waterways. The best-known surviving examples are the Alakoko Fishpond near Nawiliwili Harbor, better known as the Menehune Fishpond, and the Menehune Ditch in Waimea. Archeologists agree that the stonework at these sites is different from that of the early Polynesians. Historians and anthropologists suggest that perhaps the Menehune were a pygmy group of laborers brought from Tahiti by early Polynesians. Indeed, the word manahune, in Tahiti, means commoner. Hawaii's Menehune are said to have departed the islands forever from a point on Kauai, leaving on a three-tiered floating island. There are people still living, however, who claim to have seen Menehune in the woods and deep in the valleys of Kauai. These remnants of the once-numerous tribe are blamed for many unexplained acts of mischief.

The most prominent physical attribute of Kauai is its gentle rolling farmlands — the acres and acres of emerald-green wands of sugarcane. Sugar carpets the lowlands and creeps up the sides of the mountains. It parades down to the sea and surrounds the small towns. It fringes the opulent resorts. It is everywhere. The first of the great sugar plantations in Hawaii was started on Kauai when William Hooper, a young man from Boston, planted the first twelve acres in 1835. The land was owned by Kamehameha III and Kaikioewa, who was governor of Kauai and situated at Koloa. Hooper built Hawaii's first sugar mill the following year. His acreage produced record yields and Hooper continued to expand. He set the tone for the plantations that followed, initiating an entire way of life for the nation and ultimately leading to its downfall, for it was the Americans with sugar interests who overthrew the Hawaiian monarchy.

Hooper, in setting up the first plantation, provided his workers with houses and gardens, medical care and dairy products. He issued company money that could be spent in the company stores in the company town. Other plantation owners followed his example. The life of Kauai today is still centered in small plantation towns like Hanapepe, Hanalei, Kapaa and Koloa. Some plantation towns, like Hanapepe, have maintained their rustic, unscrubbed character, admitting perhaps a psychedelic art gallery or small restaurant to their time-worn facade. Koloa, at the opposite end of the spectrum, has been completely remodeled and rehabilitated and is capitalizing on its quaint charm. The old general store, soda works and hotel for itinerant salesmen have been replaced by boutiques, fine restaurants and a pizza parlor. Close to the Poipu resort area, the old town has been given a new, and very nice, life. The other plantation towns straddle the times. They are all wonderful vestiges of a way of life that is vanishing.

Sugar isn't all that grows well on Kauai. The island is also called the Garden Isle. It abounds in a profusion of tropical flora, bright and uninhibited in the wild and lovely, cultivated and labeled in a variety of botanical gardens.

At Christmas and lasting through the winter, the poinsettia plants are often bigger than the houses. On moonlit nights in summer, the luminous night-blooming cereus flowers in ghostly beauty across the countryside. Through the seasons, plumeria trees and wild ginger scent the air with a heady perfume.

The many gardens of Kauai, open to the public, each specialize in a different aspect of tropical beauty.

Menehune Gardens near Nawiliwili features a giant Chinese weeping banyan rising from its flowerbeds. A feature of Menehune Gardens is its exhibit of labeled medicinal plants. The ancient Hawaiian pharmacopeia drew from an amazing assortment of plants.

Olu Pua Gardens in Kalaheo is actually five great gardens in one, linked by a jungle walk. The hibiscus garden is bursting with more than 130 varieties.

Right near Poipu Beach Resort, Kiahuna Plantation Gardens numbers in its bouquet more than 4,000 species of plants, including some exceptional water lilies and the famous Moir succulent garden.

The Pacific Tropical Botanical Garden was chartered by the Congress of the United States in 1964 for the cultivation of and research into tropical plants. The beautiful garden is located at Lawai and encompasses Lawai-Kai, the former Allerton estate, noted for its formal gardens. Lawai-Kai was once the royal retreat of Queen Emma.

Kauai itself is a garden, with stands of native forest and jungles bright with riots of blossoms. There are delicate orchids as small as a fingertip and brilliant red torch ginger larger than a fist.

From the remote, rain-drenched heights of Waialeale to the sun-soaked beaches of Poipu, Kauai appears to be very close to Paradise. Hollywood is right.

LANAI

It may be known as the Pineapple Isle now, but in the old days Lanai was definitely "the wrong side of the taro patch," and had a reputation as an abode of ghosts and evil spirits. Even the early Hawaiians shunned it. For almost a thousand years after they had settled the other major islands in the archipelago, they left Lanai alone.

Kaululaau, the first settler to arrive on Lanai, didn't really want to be there — in fact, he wanted no part of the place. He was a prince, the son of Maui's King Kaalaneo. He was also a royal pain in the neck, a strapping young lad who loved to "do his own thing," which happened to be uprooting breadfruit trees. The Maui village of Lahaina, then known as Lele, was being methodically stripped of this necessary food tree. When father-son communication broke down completely, the king had Kaululaau banished to Lanai. It was a desperate move, but the father was at his wits' end and figured the noble juvenile delinquent would either shape up or be killed by the legion of demons that reportedly inhabited Lanai. The deal was that if Kaulalaau saw the error of his ways and decided to say sorry, he would simply light a fire on the beach and his father would dispatch a canoe to bring him home.

As the story goes, Kaululaau turned out to be a great ghost-buster. He hid in caves by night, when, as everyone knows, evil spirits are at their best. During daylight he stalked them, seeking out their lairs. The strength he had gained in pulling out those breadfruit trees served him well as he fought the demons, one by one, to the death. When he had killed the last of them, he lit his bonfires on the beach and his father brought him home.

To this day, there are very few breadfruit trees on Lanai, and people claim that is because Kaululaau, before he went back to Maui, went on a gleeful last spree, ripping out every breadfruit tree he could get his hands on. An officer on Captain Cook's voyage of discovery heard about Lanai's scarcity of breadfruit and duly noted the deficiency in his journal, adding, however, that "it abounds in roots such as yams, sweet potatoes and tarrow." (Many men in Cook's company saw the Hawaiian Islands simply as a pantry.)

It took the great king himself, Kamehameha I, to make the place fashionable. He built a summer home at Kaunolu Bay where he periodically retreated from the burdens of office simply to fish and enjoy the ocean. His warriors were not so lucky. To prove themselves to their commander-in-chief, they would leap into the ocean from the 62-foot cliff opposite the king's residence, hoping to clear the fifteen-foot rocky ledge at the base of the cliff. It is now called Kahekili's Jump, after the Maui king who first popularized the feat by leaping from Black Rock at Kaanapali, Maui.

Kaunolu today is an archaeologist's dream, with the remains of more than one hundred dwellings, extensive agricultural plots, Kalulu Heiau temple and even some gravestones.

One of Hawaii's most colorful characters, a New Englander named Walter Murray Gibson, founded a Mormon "kingdom" on Lanai in 1855. He was eventually excommunicated from the church for some alleged irregularities and the community was abandoned.

Ventures in ranching and sugarcane were both tried on Lanai, but neither took hold. All that remains of the latter industry is a weathered old church where the Maunalei Sugar Company town of Keomuku Village once stood. Ranching survives only in the minds of some of the older men, who spent their youth riding the range as paniolos and reporting to the ranch house at Koele. Many islanders still enjoy horseback riding and keep their animals grazing in meadows beside their homes.

The first pineapple was planted on Lanai in 1924 by James Dole, a Boston businessman turned plantation owner. He bought the entire island from missionary descendants Harry and Frank Baldwin for slightly more than a million dollars. Dole spent another two million constructing a harbor at Kaumalapau for shipping his pineapples to market and for establishing Lanai City, as he ambitiously called his plantation town. He built the town to accomodate the 150 laborers he imported to work his plantation. The new families, Japanese, Chinese and Filipino, mixed happily with the resident Hawaiian population. Today the population stands at 2,500 people.

Lanai City sits at a cool elevation of 1,650 feet above sea level, amid a setting of Norfolk Island pines and mountain scenery. The wide streets are laid out in a grid around the town square. There are no buses, no public transportation, no stoplights, no souvenir shops. There are well-attended churches, a rarely used jail, a post office where everyone gathers to collect the mail, a couple of grocery stores, a bakery, the car and jeep rental agency of the Oshira family and one hotel — the Lanai Lodge — all ten rooms of it. The only bar in town is on the porch of the hotel. It closes at ten every night.

No one needs an alarm clock because the Dole wake-up whistle rouses everyone (some say even long-departed ghosts) at five in the morning. It pierces the air again at six to announce that the trucks are leaving for the fields.

The Palawai Basin, the crater of an extinct volcano, contains some of the richest farmland in the state. Its sole crop is Dole pineapple, row upon row of it. The sweet, golden fruit with its sharp, spiky leaves has been the center of Lanai life for more than six decades.

The island's pastoral tranquility is about to be altered. Castle and Cooke, which owns Dole company and also owns 98 percent of the island, is introducing the first resort on Lanai, opening in 1989.

The first of two planned resorts sits on beautiful Hulopoe Beach, an acre of white sand smiling along the bay. Just around a point of land is Manele Bay and its fine white beach. The bay is a marine sanctuary and offers some of the finest snorkeling in the islands.

There is much awaiting the visitors who will come to the new resorts, and discover a "new" Hawaiian island.

Aside from the lovely swimming beaches, there is a wind-swept Shipwreck Beach with its treacherous reefs. The rusting hulk of a World War II military ship is mute testimony to the merciless ferocity of the wind and sea along that stretch of northeast coast. The beach is littered with timber and parts of ships. In the days when the Yankee whaling fleet called at Lahaina, Shipwreck Beach claimed many a noble vessel.

One of Hawaii's strangest geologic curiosities is the Garden of the Gods. Seven miles from Lanai City, the raw red earth of the island is exposed, like a scar that never heals. It is strewn with rocks and boulders that look as if they rained from the sky in a cosmic hailstorm or a fit of divine pique. At sunrise and sunset, the Garden of the Gods glows with rich russet tones and the ancient rocks almost hum.

Hunters will find an abundance of axis deer and such prized game birds as pheasant, quail, partridge and wild turkeys. Few sportsmen go home empty-handed.

Hikers have the fascinating Munro Trail to explore. The seven-mile loop can also be traversed by four-wheel-drive jeep. It climbs to Lanaihale, the highest point on Lanai, with views of every major Hawaiian island except Kauai and Niihau. The trail dips into misty valleys and Hookio Gulch, scene of a battle that disturbed one of Kamehameha's Lanai holidays in 1778. The flora along the way is verdant green, sparked with brilliant wildflowers and sprinkled with dew.

The trail is named for a New Zealander, George C. Munro, who managed the Lanai Ranch in the early twentieth century. Munro was a naturalist by hobby, and when he rode about the mountains on horseback he always carried a sack of seeds on his saddle. He'd sow them as he rode. His most successful introduction was the stately Norfolk Island pines that have not only greatly aided Lanai's ground water system, but have come to be associated with Lanai, as palms are with other islands.

Lanai is poised on the edge of time, about to trade one way of life for another. It is expected that the introduction of resorts will increase the island population to 15,000, so that newcomers will vastly outnumber the old-time families. Not everyone is happy, but watching the pineapple fields shrink in size as cheaper, foreign-grown fruit seizes more of the market, and saying good-bye to their children as they move away to seek education and jobs, has convinced most of Lanai's people that the change is necessary, and that maybe sharing their lovely Lanai won't be so bad, after all.

MOLOKAI

Only twelve jet minutes from the bright lights of Wakiki are the dark, brooding hills of Molokai, with barely a light burning after sundown.

This little-known island is the closest in distance to urban Oahu, and probably the farthest in spirit. With the exception of closed Niihau, Molokai is the only Hawaiian island where native Hawaiians are the majority of the population, which numbers about six thousand people.

Molokai is small, 37 miles long and about ten miles wide. Its highest point is the summit of Mauna Kamakou, rising almost five thousand feet.

In the beginning, Molokai was a part of one large island, which geologists call Maui Nui. Maui Nui encompassed what are now the separate islands of Maui, Molokai, Lanai and Kahoolawe.

Molokai's distinctly Hawaiian flavor is a reflection of the very real spirit of aloha, cherished and practiced by the people of the island. It is this spirit that has earned Molokai its appellation, the Friendly Island.

Remnants of the Hawaiian culture are everywhere, in the ruined temples all over the island, among the abandoned stone structures in the jungles of Halawa Valley, in the petroglyphs of Kawela and, most prominently, in the fishponds that ring the island's shores. These ancient ponds enabled the Hawaiians to maintain a steady supply of seafood. In a highly technical process that is just beginning to be understood by modern scientists, the Hawaiians raised fish and maintained them, much the way modern ranchers maintain animal herds. Most of the ponds are no longer serviceable due to neglect, the effects of silting and tidal action. There are, however, a few working ponds, and some that have been resurrected by marine scientists for research.

Molokai has only one town of consequence, Kaunakakai, and it's not big enough to need a traffic light. With its stores and simple restaurants, Kaunakakai resembles, more than anything, a set from an old "Gunsmoke" episode.

There were probably more people living in Halawa Valley in ancient times, tending their taro and other crops, than there are now in Kaunakakai.

Halawa is the end of the road of East Molokai. The valley cuts deeply into the rugged north shore mountains. At its head are two waterfalls, Hipuapua thundering with a 500-foot cascade and half-as-high Moaula, which plunges into a pond said to be inhabited by a capricious lizard goddess, or mo'o. Swimmers are advised to toss a leaf into the pond before entering. If the leaf floats, the mo'o is having a bad day and it is best to let her have the pool to herself.

Molokai abounds in legends that have become shrouded in time. The kahunas, or priests of the island, were said to have been so powerful in their sorcery that they were more feared than warriors, which is just as well, because Molokai was too small to raise much of an army.

It is a wonder that Molokai is not overrun with visitors, as some of Hawaii's greatest attractions are here. One of the longest and widest beaches in the Islands, Popohaku, lazes for two and a half miles along the western end, and most days it is deserted.

The highest sea cliffs in the world parade along the northern shore, sheltering deep, lush valleys where some Hawaiian families have retreated from the twentieth century, to live as their ancestors did for centuries, subsisting on the bounty of the land and the sea. It is a region of waterfalls, offshore isles and breathtaking beauty.

Anchoring one end of this green fortress is Makanalua Peninsula, better known as Kalaupapa, the place of exile for victims of Hansen's disease, or leprosy. Here Father Damien labored among the sick until he himself succumbed to the dreaded disease in 1890. When he arrived at Kalaupapa, he had the materials to build a church and a home for himself. He built the church and invited the sick to sleep there. Lacking homes, they all had been sleeping in caves. Damien slept under a tree until each of his flock had a roof over his head. There are more graves than homes at Kalaupapa, which sits at the foot of a 2,000-foot cliff, accessible only by foot or mule traversing a three-mile switchback trail. A small airport services the colony and a barge brings supplies every few months. Although Hansen's disease is now curable, some of its victims, who suffered the ravages of the disease before a cure was known, choose to stay in isolation at Kalaupapa, in an incredibly beautiful setting, with the fellow sufferers they have known all their lives. Saint Philomena's Church, built by Father Damien, is there, as is the famous Siloama Church, or Church of the Healing Spring.

A guided mule ride down the cliff trail, with its twenty-six switchbacks, takes visitors from "topside" to Kalaupapa, where patients conduct a tour of the peninsula.

Wild hoofed African animals, more than four hundred of them, have made Molokai their home. Originally imported to eat the kiawe scrub that plagues ranchers by encroaching on their pastureland, the animals have increased in number to the point where Molokai Ranch, which started the program, earns extra income selling the offspring to zoos around the world. Camera safaris to photograph the rare impala. oryx, ibex and other animals are conducted through the 800-acre wildlfe reserve.

Common cattle have not fared as well on Molokai as the exotic species. In an effort to eradicate bovine tuberculosis, which has periodically cropped up in Molokai herds, the Hawaii State Department of Health in 1986 ordered the slaughter of all cattle on the island. The larger ranches, such as Molokai Ranch, were able to absorb the shock, but smaller ranchers suffered great economic, and in some cases, emotional hardship.

Molokai's other amazing surprise is the imposing six-foot phallic rock, Kaule o Nanahoa, at Palauu State Park. Legend holds that any infertile woman who sleeps by the rock will become fertile.

In the uplands of Moloka's mountains, the Nature Conservancy of Hawaii maintains the Kamakou Preserve, a vast, forested area of sandalwood, wild plum and native Hawaiian plants. Many varieties are on the national register of rare and endangered plants. Monthly nature hikes are conducted in the preserve.

Pineapple production has been on the decline for years on Molokai, with Dole completely shut down and Del Monte having announced its intention to phase out its pineapple operation. There is still some diversified agriculture, with corn being a primary crop.

Most of Molokai's people live in rural settings or in small plantation towns like Maunaloa and Kualapuu. Their small, wood-frame homes with tin roofs are almost overpowered by their gardens, bursting with flowers and overhung with giant mango, banana and lychee trees, ripe with luscious fruit.

These people have managed to maintain their traditional way of life in spite of their proximity to Oahu, even though Honolulu's city lights twinkle seductively across the Pailolo channel, only twenty-five miles away.

No alien land in all the world has any deep, strong charm for me, but that one; no other land could so longingly and beseechingly haunt my sleeping and waking through half a lifetime, as that one had done. Other things leave me, but it abides.

—Mark Twain

A traditional Hawaiian grass house and its inhabitants at Moanalua, Oahu, around 1890.

Dancers parade with tiki torches at a festive luau— the traditional Hawaiian feast.

Hawaiian lei sellers around 1900. Even then, Hawaii was ethnically diverse, as the sign for a Chinese tailor shows.

Native Hawaiian hula dancers perform at King Kalakaua's jubilee celebration on the grounds of Iolani Palace, Honolulu, in 1886.

King Kalakua (at center) and his staff on the steps of the Iolani Palace in Honolulu, around 1886.

Queen Liliuokalani, the last Hawaiian monarch. She succeeded her brother King Kalakua in 1891 and reigned for two years.

King Kalakua and writer Robert Louis Stevenson, some time in the 1880s.

Pa'u riders, dressed in the riding habit of the monarchy era, sweep across the meadows near Kamuela on the island of Hawaii.

The people of Hawaii are ethnically diverse, but some are of pure Polynesian descent. Harry Pahukoa and his family, pure Hawaiians, live at Keanae on Maui.

In the Waimea-Kamuela district at the foot of Mauna Kea on the island of Hawaii, these pa'u riders wear the white riding outfit of the 1800s, when Hawaii was ruled by a monarchy.

Warmth and hospitality characterize the people of Hawaii. Visitors can count on a friendly and warm welcome.

Leis, the traditional flower garlands of the Hawaiian people, are a symbol of peace and friendship. Natives and visitors alike wear them whenever possible.

A tree bursting with plumeria blossoms and a pretty girl—a classic vision of the delightful Hawaiian islands.

Dancers wearing traditional ti-leaf skirts whirl in a performance on Oahu, the Gathering Place.

The Kodak Hula Show at Waikiki on the island of Oahu has been attracting visitors for more than fifty years.

Tahitian dance skirts are brightly colored and beautifully detailed.

Tahitian dancers move rhythmically to the beat of the drums at a Polynesian revue.

Musicians entertain at the Polynesian Cultural Center at Laie on the north shore of the island of Oahu. Native villages representing the cultures of Fiji, Tonga, New Zealand, Tahiti, Samoa, the Marquesas and Hawaii are found here.

The Pageant of the Long Canoes is a daily (except Sunday) extravaganza held at the Polynesian Cultural Center at Laie. Most of the performers are students at the Hawaii branch of Brigham Young University.

A steaming pig, cooked to perfection, is brought from its imu oven in time to be the main course at an evening luau at Kailua-Kona on the island of Hawaii.

Poi, a staple food of the native Hawaiians, is made from the pounded root of the taro plant, as this man demonstrates at the Polynesian Cultural Center at Laie, Oahu.

The Hawaiian art of lauhala weaving is demonstrated by a woman at the International Market Place at Waikiki on Oahu.

Jade Bernades, a native of Maui, the Valley Isle, is framed by the giant leaves of a taro plant.

On Kauai, the Garden Isle, a worker uses a sharpened spike to tear the outer husks off coconuts.

Wailani Johansen, a quilter on Maui, exhibits a traditional Hawaiian pattern.

Honolulu artist David Asherman works with Hawaiian themes.

Native fishermen gather along the shoreline of Kauai. The interior of Kauai is mountainous and rugged, but the eastern side of the island has long, white beaches.

Traditonal Hawaiian skills, such as throw-net fishing, are demonstrated at Puuhonua National Historical Park, on the Kona Coast of the island of Hawaii.

Fearsome ki'i or tiki images are symbols of the ancient Hawaiian religion. Many tikis and ancient temples can still be seen today.

Tikis stand guard at Puuhonua o Honaunau, the Place of Refuge, on the island of Hawaii near Kealakekua Bay. In ancient times, warriors defeated in battle raced to this area for sanctuary from the victors.

An outrigger canoe waits to take visitors surfing on the long rollers at fabled Waikiki Beach on the island of Oahu.

The twin bows of a beached double-hulled traditional Polynesian canoe are draped in leis at Makaha, Oahu.

The champion paddlers of Oahu's Outrigger Canoe Club demonstrate their winning style. The outrigger, which gives the canoe additional stability, can be seen at right.

Outrigger canoes brought the first Polynesians to the island of Hawaii around 750—more than a thousand years ago. Today's Hawaiians still enjoy racing their canoes in the many regattas and competitions held each year.

A carved tiki on the bow of an outrigger canoe is a reminder of ancient Hawaii on urbanized Waikiki Beach, Oahu.

Following Page:
A team of canoers is only a speck on the ocean at sunset. Although today's canoers are rarely out of sight of land, the ancient Polynesians who discovered and settled Hawaii traveled thousands of miles across the open ocean in their canoes.

Old sugar mills are part of the rural Hawaiian landscape, especially on Maui, the Valley Isle.

Sugar cane carpets Maui's Central Valley in this aerial view. Hawaii has the most advanced sugar technology of any area in the world.

Sugar cane, a major export crop valued at nearly $400 million a year, grows tall and green in the ideal climate of Maui's Central Valley.

Stalks of sugar cane tower above the harvester who cuts them down. Despite advanced technology, many sugar cane operations must be done by hand.

A sugar worker ignites the cane for carefully controlled burning—an essential part of the harvest.

The smoke from burning sugar cane drifts over a field at Waialua on the island of Oahu.

A taro farmer takes a break on Maui's Keanae Peninsula.

Taro and banana patches grow side by side at Keanae Peninsula on Maui. The large, starchy root of the taro plant is used to make poi, the staple food of the ancient Hawaiians.

The Keanae Peninsula of Maui is famed for its extensive network of carefully irrigated taro patches.

Taro patches form a lush green quilt in this aerial view of Maui's Keanae Peninsula.

Pineapple must still be harvested by hand. Note the heavy gloves worn by this worker to protect his hands against the sharp spikes on the pineapple's stalk and skin.

Pineapple being harvested on Molokai, the Friendly Isle.

In all Hawaii, more than 35,000 acres are devoted to pineapples.

Hawaiian pineapple is world-famous for its quality—sweet, juicy, and delicious.

Harvesting pineapples is only partially mechanized.

Much of the work must still be done by hand.

Hawaii's pineapple crop is valued at more than $250 million each year.

Ranch lands comprise 52 percent of the total land area of the state of Hawaii. These horses graze at a ranch on Oahu.

Horses graze the rich pastures of the Parker Ranch at Waimea on the island of Hawaii. The Parker Ranch, at the foot of Mauna Kea, is the second-largest cattle ranch under the American flag.

A tunnel of ohia trees frames these cattle at a pasture on Hawaii, the Big Island.

These horses are corraled at the Kualoa Ranch on Oahu. The beautiful Kaawa Valley is part of the ranch.

The upland ranch country of the Hana region on the southeast coast of Maui is ideal grazing and farming land.

A red Angus bull is lord of all he surveys on the Parker Ranch at Waimea on the island of Hawaii.

When most parts of the American West were still unsettled, the Parker Ranch was already a going concern.

Ignored by the cattle grazing in their pasture, a rainbow and its faint double arch over a ranch in Maui's upcountry.

A dramatic lava rock formation makes this sand trap a particularly hazardous one for the hapless golfer.

Cooled streams of molten rock from lava flows are a natural hazard at the Mauna Kea links on the island of Hawaii.

The Royal Kaanapali golf courses are a major attraction of the Kaanapali Beach resort complex on Maui.

Golden light bathes a golf course on Maui. The state of Hawaii is a golfer's paradise, with many spectacularly beautiful and challenging courses.

The greens at Princeville on Kauai, the Garden Isle, are among the most scenic in the world.

Swimmers leap into the ocean from a rock at Waimea Bay on the island of Oahu. The spectacular beach here is excellent for picnics, but the surf often reaches heights of well over six feet.

Two backpackers enter the moonscape world of Haleakala Crater on Maui. From the crater's topmost rim to its floor is a drop of 3,000 feet; the floor measures 25 square miles.

Shooting across the waves on jet skis is a popular sport, especially in the waters around Waikiki.

A young surfer tries a wave at Waikiki. Surfing, whether for beginners or experts, is a top attraction here.

Waimea Bay on the northwest coast of Oahu is one of Hawaii's best surfing spots, combining a magnificent beach with towering waves.

Surfer Hans Hedemann rides the tube of a wave—he stays on the crest of the wave as it breaks behind him.

Expert surfers ride the mammoth breakers found during the winter months on the north shore of Oahu island.

Wipeout! Even expert surfers are often fooled by the untameable power of the ocean.

A surfer calls it a day and heads home from a favorite spot on the north shore of Oahu.

Windsurfers from around the world come to test their skills against the wind and waves of Hawaiian waters.

In the calm of early morning sailboards wait for their users on a beach at Maui.

A windsurfer leaps over the waves. The sport of windsurfing, combining the challenges of sailing and surfing, is fast becoming a favorite among visitors to Hawaii.

The rider of this colorful windsurfer is almost hidden by the sail as he races across the water.

Jim Ingham "flies a hull" on his Hobie catamaran in Kailua Bay, Oahu.

This rainbow-sailed catamaran at Waikiki on Hawaii island shows one of the advantages of these double-hulled boats—they can be drawn up directly onto the beach.

The trimaran Milolii plies the waters off Wailea on the island of Maui.

These boats have three hulls and are virtually unsinkable.

A bright yellow umbrella, a dazzling white beach, a cerulean blue sky and an ultramarine ocean make up this classic view of a perfect Hawaiian beach.

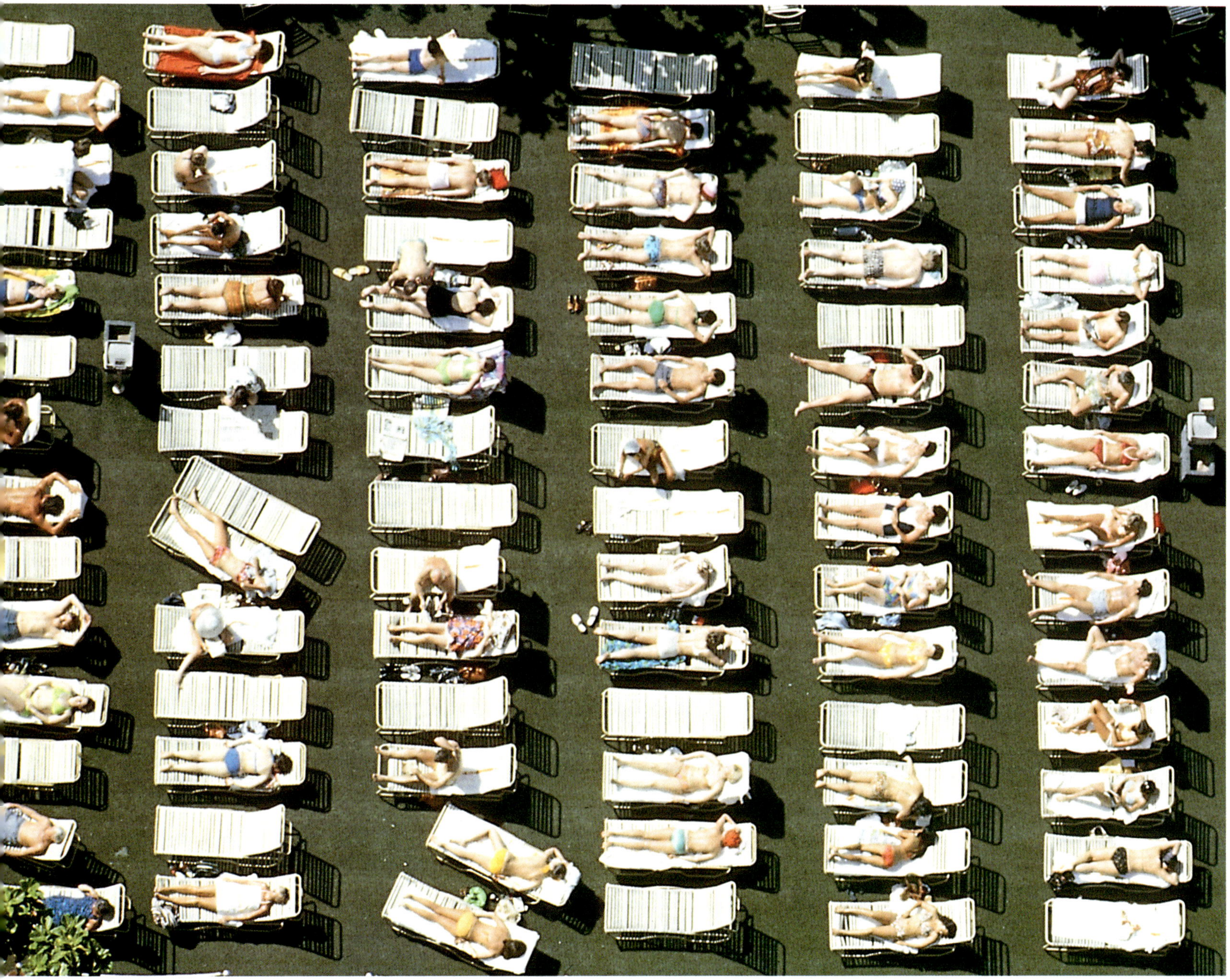

The lounges at any Hawaiian resort are always filled with appreciative sunbathers enjoying the warm sun.

Tourism is an important industry in Hawaii, and Hawaiians know how to please their visitors. Loungers around the sparkling pool at this resort can enjoy the sun, the water, and all the comforts of the hotel.

Competitors enter the water for the swimming phase of a triathlon at Waikiki. As part of the race they also ride bicycles and run a marathon.

The Kona Coast of the Big Island of Hawaii is famous for record marlin catches. Sport fishermen by the thousands come each year to try their luck.

A snorkeler floats quietly through a reef, enjoying his private view of an underwater world.

A school of fish engulfs a visitor to their reef.

This snorkeler is exploring in Lydgate Park on the island of Kauai.

Yellow tangs are a common fish in Hawaiian waters. They are often encountered by snorkelers and divers.

One of the best places to observe Hawaiian marine life without getting wet is the natural reef tank at Sea Life Park at Makapuu on the island of Oahu.

Fantail filefish float majestically near a coral reef.

A wrasse, a small fish commonly found around Hawaiian reefs, hovers near a piece of coral.

A red hawkfish displays its spiny dorsal crest
as it rests on the ocean floor of a coral reef.

The ocean waters around Hawaii abound in edible seafood, including shellfish, crustaceans and fish of every sort. Seafood is found on the menus of restaurants throughout the Islands.

Koi (carp) swim among hibiscus blossoms in an ornamental fishpond. These beautiful fish thrive in Hawaii, but they were brought to the Islands from the Orient.

Prize koi (carp) are an attraction of the Japanese gardens at the Byodo-In Temple at Haiku on Oahu.

Although not native to Hawaii, mongooses thrive in their adopted land—as this alert colony shows.

Mongooses were introduced to Hawaii to help control the rats that were plaguing the sugar cane growers. However, since the mongoose is active during the day and the rat is active at night, the two rarely meet.

The Brazilian cardinal is an introduced species that thrives in Hawaii. The Islands, with their incredible variety of avian life, are a favorite destination for birdwatchers.

A peacock surveys his domain at Waimea Falls Park on the island of Oahu. The 1,800 acres of the park are a sanctuary for tropical birds and plants.

The giant floating leaves of Amazon lily pads will hold a 60-pound child afloat.

Tropical species from other parts of the world thrive in Hawaii.

A bud of the beautiful plumeria flower promises a spectacular, fragrant blossom in a few days.

The fleshy, spiny leaves of the bird of paradise plant are a surprising contrast to the lovely blossom.

The Hawaiian climate is ideally suited for raising exotic flowers such as this heliconia.

The export of exotic flowers is an increasingly important business in Hawaii.

Anthurium farms abound in the Hilo region of the island of Hawaii.

This unusual flower is exported all over the world.

Orchids, so difficult to grow on the mainland, thrive in the Hawaiian climate. Orchid farms prosper in the Hilo area of the island of Hawaii.

The Nani Mau Gardens at Hilo on the island of Hawaii is famous for its displays of exotic orchids.

Cattlya orchids are a specialty of Nani Maui Gardens at Hilo.

The hibiscus is the state flower of Hawaii. It comes in hundreds of varieties, with many different colors but always a delightful aroma.

The hibiscus is traditionally used to make leis, the Polynesian flower garlands customarily given as tokens of affection and welcome.

More than five thousand types of hybrid hibiscus, in a rainbow of colors and combinations, are known.

The state tree of Hawaii is the kukui, also known as the candlenut tree. The nuts of this tree provided ancient Hawaiians with oil for lights and medicine.

A graceful lychee tree awaits its season to flower and fruit.

The sap found in the spiny fronds of the succulent aloe plant is good for soothing burns and chapped skin.

Dew sparkles on the fronds of a palm tree. These graceful trees are found throughout the Islands.

In the lush climate of the Hawaiin islands bananas grow wild for the picking.

A potpourri of fruits and vegetables grown in Hawaii—including tropical fruits such as pineapple, guava, bananas, and coconut.

The fronds of a hala tree are silhouetted against the evening sky at a beach near Honolulu.

A cluster of coconut palms. The graceful, curving trunks and leafy fronds of these tropical trees are a common sight in the Islands.

Tall coconut palms lean into the sunset. These graceful symbols of the tropics are found on many Hawaiian beaches.

The idyllic east coast of Maui has areas of tropical rainforest.

This side of the Valley Isle offers some of Hawaii's finest scenery.

Although visitors often think of Hawaii as a jungle rainforest fringed by beaches,

in fact many of the islands have long vistas of rolling grassland dotted by stands of trees.

A stream called Kepaniwai tumbles through the Iao Valley on Maui.

It was here that King Kamehameha won his battle for control of the island.

Lush rainforest vegetation and rushing mountain streams characterize Maui's Hana Coast. Here ferns and a giant philodendron hang over a tiny stream.

Rainbow Falls at Hilo on the Big Island of Hawaii crashes over a cliff into a pool framed by wild red ginger.

Waikani Falls tumbles down a mountainside into the fern-choked forest near Hana on the island of Maui.

The water is so soft and pure in the mountain pools of Maui that it is the world's best hair conditioner. Delightful waterfalls abound on the east coast of the island.

The Hana district on the east coast of Maui is immensely varied and scenic. Spectacular Wailua Falls cascades through the tropical forest to continue down the mountainside.

Keawanui Waterfall tumbles through a deep gorge on Molokai, the Friendly Isle.

Akaka Falls at Akaka Falls State Park, north of Hilo on the island of Hawaii, plunges over 420 feet in a sheer drop over a volcanic cliff.

The famous black sand beaches on the island of Hawaii were formed over the eons by the pulverizing action of the ocean waves on the lava shoreline.

Falling coconuts are a posted hazard at a black sand beach on the Big Island of Hawaii.

The southeastern shore of the island of Hawaii has two world-famous black sand beaches: Kaimu and Punaluu.

Makapuu Beach is pounded by surf. The lava-strewn shoreline is guarded by Makapuu Light high on the cliffs above.

Surf sweeps ashore at Kauai, the Garden Isle. Although the world's wettest spot, Mount Waileale, is on the island, the beaches are warm and sunny.

The waves pounding against the shore of the island of Maui may be discouraging to swimmers, but they are a surfer's delight.

The virtually inaccessible but breathtakingly beautiful Na Pali coast of Kauai can be seen by taking one of the popular helicopter tours.

The Chinaman's Hat, an offshore isle, is a distinctive landmark off the windward coast of Oahu.

Hanauma Bay in Koko Head Park on the island of Oahu is a marine-life sanctuary and a popular snorkeling spot.

Sushi Island stands out against the backdrop of Molokai's North Shore cliffs.

Ocean mists often shroud the towering sea cliffs of Molokai's mysterious North Shore.

Waves send a geyser of salt water up through the Halona Blow Hole on the black lava cliffs of the coast of Oahu.

Centuries ago a stream of molten lava flowed into the sea off the coast of Maui island.

The powerful action of the waves has eroded the formations into these ethereal shapes.

The Na Pali coast of Kauai as seen from the air. Geographically and historically, Kauai is the oldest of the islands in the Hawaiian chain.

The North Shore of Molokai, Hawaii's fifth-largest island, has the world's highest sea cliffs.

The towering slopes of the West Maui Mountains are all part of a single volcano, Mauna Kahala-wai.

The emerald cliffs of Kauai's rugged Na Pali coast on the southern shore of the island plunge 4,000 feet to the surf below.

The rugged, lushly forested Iao Valley is one of the scenic wonders of Maui. This tranquil park was the scene of a bloody battle in 1790 when King Kamehameha conquered the island in the famed Battle of Kepaniwai.

Following Page:

Clouds gather near the peaks of the Koolau Mountains, which tower over Kualoa Point on the island of Oahu.

The Garden Isle of Kauai is blessed with beautiful beaches. This one is at Wainiha Bay.

A rainbow spans Waimea Canyon on the south shore of Kauai. Often called the Grand Canyon of the Pacific, its depth and breadth are awesome.

The sunrises and sunsets on Maui are spectacular blazes of flaming colors.

The sky above Hilo on the Big Island of Hawaii is ablaze with color as the sun rises for another magnificent tropical day.

Hawaiian sunrises and sunsets are among the world's most dramatic. Stunning views can be seen from almost anywhere.

The beach at Makena is one of Maui's best—a wide, uncrowded stretch of soft sand, gentle surf and great natural beauty make it a favorite.

The Makanalua Peninsula on Molokai is the site of the Kalaupapa settlement for sufferers from Hansen's disease (leprosy). Established by Father Damien in 1873, the colony can be reached only by air, on foot, or by donkey down a steep, 2,000-foot mountain trail.

The strange lava formations along the Keanae shoreline on the island of Maui are constantly battered by the pounding surf.

Honolulu has been a port of call since before 1800. At first whalers and traders were the only visitors, but now cruise ships and pleasure craft fill the harbor. Diamond Head Light guides boats safely into the port.

Diamond Head Light is the brightest lighthouse in the Pacific. It stands sentinel on the Oahu shoreline near Honolulu.

The beacon of Diamond Head Light beams out over the ocean as the sun sets.

On the north shore of Kauai, the Garden Isle, is the Kilauea Lighthouse. The lighthouse is open to visitors; the isolated promontory on which it stands is a refuge for nesting seabirds.

Hulihee Palace, in the village of Kailua-Kona on the island of Hawaii, was once the summer retreat of the Hawaiian royalty.

The 900-year-old Byodo-In Temple in Japan is duplicated in exact detail at the Valley of the Temples Memorial Park, beneath the majestic cliffs of the Koolau mountains on the island of Oahu.

Royal palms flank the entrance to a Hawaiian church. The first American missionaries arrived on Hawaii from New England in 1820. Today Hawaii's religions are as diverse as its cultural heritage.

Kaahumanu Church, set among the sugar cane fields, is one of the outstanding landmarks at Wailuku on the island of Maui.

The exterior of Saint Philomena's Church, at Kalaupapa on Molokai. It was built by Father Damien, the Belgian who came for a few weeks and stayed for the rest of his life.

The interior of Saint Philomena's Church. The simple wooden church stands near the settlement's old cemetery, where Father Damien is buried.

The United Methodist church calls the faithful of Kauai. The first Methodist missionaries arrived in Hawaii in 1854.

The historic Waioli Mission church still stands at Hanalei near Princeville on the island of Kauai.

Built in 1834, the church and mission house have been restored by descendants of the first missionaries.

The Sheraton Maui, at Black Rock, anchors one end of Kaanapali Beach, famed for its luxurious resorts.

Luxury condominiums cluster around a pool at the exclusive Kapalua Resort on Maui.

The high-rise towers of downtown Honolulu glow in the sunset. Honolulu has been the capital of the Hawaiian islands since 1845.

The Ala Wai Yacht Harbor is home to many of the pleasure boats on Oahu.

The famous statue of King Kamehameha stands in front of Aliiolani Hale (the judiciary building) in downtown Honolulu.

Downtown Honolulu is a study in contrasts: sleek, modern office towers rise above older buildings from Hawaii's territorial past.

The Ala Wai Yacht harbor at Honolulu is home to two yacht clubs and a large flotilla of pleasure craft.

Pleasure sailing is a favorite pastime for both natives and visitors to the Islands. Marinas and sheltered harbors abound; this one is in Honolulu.

A full moon rises above the resort towers of Waikiki on the island of Oahu.

Diamond Head is the backdrop of the Ala Wai golf course and the high-rises of Waikiki.

Waikiki Beach on the island of Oahu is Hawaii's premier resort area—

the magnificent beach in lined by hotels providing their guests with every comfort.

The Arizona Memorial in Pearl Harbor is the state's number-one tourist attraction. The memorial commemorates those killed in the attack on Pearl Harbor on December 7, 1941.

A roster listing the names of those who lost their lives during the Japanese attack on Pearl Harbor is on view at the Arizona Memorial. There were nearly 4,000 casualties from the attack.

Visitors at the Arizona Memorial view the rusting hulk of the USS Arizona, sunk during the surprise bombing attack on Pearl Harbor.

Pearl Harbor remains an important link in America's naval defenses. The headquarters for the Commander in Chief Pacific (CINCPAC) is located on Oahu.

The American and Hawaiian flags blow in the tradewinds. Hawaii was annexed to the United States in 1898; it became a territory in 1900 and a state in 1959.

The submarine USS Bowfin, *based at Pearl Harbor, is open to visitors.*

A river of molten lava carves its way across the mountain at Kilauea in Hawaii Volcanoes National Park on the island of Hawaii.

Night views of the erupting Kilauea volcano are the most spectacular. Kilauea is the home of Pele, the legendary goddess of fire.

Kilauea spouts flames and spews forth rivers of molten lava from the earth's core. Mauna Loa is the world's most active volcano.

The island of Hawaii is only 76 by 93 miles, but lava flows from Kilauea constantly add to the 266 miles of coastline.

Even amid the desolation of a cooled lava flow, new life springs up in Hawaii Volcanoes National Park.

A vast field of cooled lava forms a dramatic wasteland at Kilauea on Hawaii island.

Trees and sulfurous steam drifting up from deep fissures in the scorched earth make a ghostly landscape at Kilauea in Hawaii Volcanoes National Park.

A hike along Devastation Trail in Hawaii Volcanoes National Park is a walk on the weird side.

Only the charred skeletons of trees remain to hint at the vegetation covered by the lava flow.

Devastation Trail takes visitors across a wasted land.

The crater of Maui's Halaeakala Volcano resembles the face of the moon so much that astronauts are trained here.

Snow crowns the heights of Mauna Kea on the island of Hawaii. At its highest the mountain reaches 13,796 feet—making it one of the world's tallest.

The extinct volcano that is Koko Head is found on the southern coast of the island of Oahu, not far from its more famous sibling, Diamond Head.

The Big Island of Hawaii is twice as large as the other islands in the Hawaiian chain. The island has a fantastic variety of terrain, ranging from stunning beaches to two of the world's highest mountains.

Clouds of sulfurous mist drift through Hawaii Volcanoes National Park, where the lush vegetation is always in danger of obliteration by a lava flow.

A beautiful Hawaiian rainbow arches across sugar cane fields near Lahaina on the island of Maui.

INDEX OF PHOTOGRAPHERS